Published by

JENNA PUBLISHING

PAT MANNING
29, Birchwood Ave,
Beckenham,
Kent, BR3 3PY

In memory of my joyful dog Jenna who walked the woods with me

JENNA

I have loved all my dogs but my Jenna was special,
Bright and trusting, listening, questioning, faithful.
Wherever I was so this little dog would be
My Jenna who walked through the woods with me.

A small fair scrap this little dog I chose her
To be mine forever it seemed that September.
Just retired the years stretched ahead invitingly
For me and my Jenna who walked the woods with me.

She'd run on ahead but then always looked round.
She'd jump and race so her legs left the ground.
She'd carry long sticks with great pride and I'd see
My dear Jenna who walked through the woods with me.

That morning she chased after a fox in the park.
She managed the hedge and even a bark.
I didn't know as I waited by the tree
That it was the last time my Jenna walked with me.

No more down the stairs tumbling to greet me
Carolling joyfully so pleased to see me
Waiting expectantly knowing she'd soon be
Out with Ben walking the woods with me.

I'm caught unawares, I think she's in her chair
Or perhaps she's sitting on her favourite stair.
She's not there and I know now that she'll never be
Any more walking through the woods with me.

                                        Be free sweet Jenna, be free.

RIVERS REMEMBERED (The River Pool)

THE BEGINNING

We sat on the veranda watching the riverbank intently. The storm had passed but we knew that the river was running dangerously high. Suddenly we saw it! A brimming-over converged into a thin silver stream running down the bank and formed an ever-growing pool of water in our field. In spite of the railway sleepers that our father had used to shore up the bend, the river pushed its left bank aside with contemptuous ease and the powerful current flooded through the 10ft gap. We jumped for joy as the water crept nearer and nearer to the pavilion. The field had become our lake to explore in the punt with a dog seated at either end enjoying the fun too.

On quieter days, we paddled on the sandy riverbed with the water rippling a few inches over our feet. In July and August, the banks enclosed us with their towering growth of the Indian plant, *Impatiens glandulifera,* imported over 150 years ago from the Himalayas. To us, it was firemen's hats. As we pushed among its thick red stems and handsome flowers from palest pink to deep claret, we came under fire from its snapping seed capsules, fascinating to pinch.

In those days we were supposed to wear funny rubber slippers when paddling. They were very difficult to pull on and off and the sand felt so fine under bare feet. I learnt my lesson one day though when I stepped on broken glass and nearly sliced off a toe. Of course, there were always the leeches.

In the corner of the field, we had a knotted rope hanging from an oak tree on which we could swing across the river. A brood of Black Leghorn chicks raised by my pet hen soon discovered another way. They took off and landed safely on the other side leaving their more portly parent clucking anxiously until their return.

We first came to the ground in 1930 because the Japanese of the Yokohama Specie Bank had just employed our father as steward and groundsman. There was a white shed at the bottom of the field, perfect for the pony, donkey, chickens and turkeys although the first heavy rain demonstrated that ducks would have been more suitable. Our father was woken up by the sorrowful braying of the donkey just managing to keep his head out of the water. The River Pool had reclaimed its flood plain!

The ground was originally part of the eight-acre field of Copers Cope farm. It was at the limit of the farm's land close to the Lewisham border.

When purchased by the Southwark timber merchant, John Cator, in the late 1700's, the intention had been to develop a high-class suburb with large houses and tree-lined roads. The Cator family leased the land out for building when the railways were built in the 1850's. Stations at New Beckenham and Lower Sydenham opened up the area. In 1961, the lighting at Lower Sydenham station was only then being changed from gas to electric. As a child I had always envied the porter who walked from light to light pulling down the chain to light the lamp with a ring on the end of a stick.

I crossed over the bridge four times a day to go to school at Haseltine Rd in Bell Green. If I had a penny, I might put it in the BeechNut chewing gum machine where every fourth time two packs were delivered. I did not really like chewing gum but it was the excitement of seeing two packs come out instead of one!

Although Worsley Bridge road provided access to the eight-acre field, housing was restricted to the higher ground in Copers Cope Rd. Building began opposite the end of Brackley Rd near the new church of St Paul. On either side of the river, the London banks used the land for their sports grounds and our field was leased to the Yokohama Bank with the National Provincial bank taking the 13-acre field over the other side. When it rained, the river overflowed, mostly our way.

There is evidence that the river existed in Roman times. Parts of the Roman road between London and Lewes lie on a line of flat land on the well-drained Blackheath pebble beds between the thickly forested London clay on the high ground to the East and the often marshy Pool valley to the West. Between the end of Greycot Road and Meadow Close, in the 7-acre field of Copers Cope farm, the Roman road is only just over a foot below the present surface. It consists of 30ft wide gravel, 11ins thick, with rut marks, resting on pebbles and flints. In this way the Romans on their way North avoided climbing up to Stumps hill but kept away from the river until a crossing over a firm river gravel near Catford.[1]

THE BEGINNING OF THE END

In 1940, two silver and blue bombs angled through the air from a German bomber aiming for the railway line where it crossed the river. One bomb fell in our field. My brother, waiting to join the RAF, was working at the factory, John Bells, painting the air raid shelter doorframes yellow. His foreman made the dry remark "That should improve the drainage". The second bomb hit its target and stones from the railway went through a line of white towels that my mother had just pegged on the line.

She did not let us forget it because we had no electricity or gas and she had spent the best part of the day boiling the laundry up in a copper outside.

When the Japanese attacked Pearl Harbour, 7.12.1941, we moved to one of the end houses in Copers Cope Rd but my dog, Rover (1936-1954), found a way back that became his daily routine. A fallen willow tree formed a bridge over the river. Every afternoon, Rover took his diagonal route across the National Provincial Sports Ground to visit his birthplace. His path passed between the wickets of the first team cricket pitch and play stopped at three o'clock on Saturday afternoons to let him take his constitutional.

After WW2, the ground was taken over by Thomas Meadows and efforts were made to stem the R. Pool by building up the land 8ft. This buried the elegant semicircular red brick steps of the pagoda-style pavilion but prevented the flooding. Today, visitors to boot sales at the Footzie Club can be barely aware of the river of my childhood, flowing as it does now deep down between the banks. The veranda has been removed and the building has lost its character.

Now that the river had lost its flood plain, it exaggerated the flooding of the Chaffinch Brook at Clockhouse.

Many alive today will remember the rainfall of Friday night 15.9.1958 when the flooding of our local streams fed by the rain in the Shirley hills caused the Beck to pour out of the Kelsey Park gates like a bore 4ft high. Clockhouse station was flooded twice; first by the Chaffinch coming from the direction of the up line and then by a tidal wave from Elmers End. Large lakes appeared all over the borough: - Shortlands, New Beckenham, Birkbeck, Lower Sydenham, Worsley Bridge, Rising Sun, White Hart at West Wickham and Sparrows Den where the downpour had revealed the forgotten streambed of the Bourne. This appeared again in February 2001 and formed a lake in Sparrows Den.

The electric train service ceased through Clockhouse station until a steam train was used to pull the coaches along. The whole of the Elmers End Green was covered and the Churchfield allotments were inundated. The Borough stream (Pool) had swept through the Reddons Rd gardens, flooding cellars forcing up the floorboards in the houses. Mr Nash, the caretaker at the Girls' Grammar School in Lennard Rd (now Cator Park School), was kept busy trying to control the 2ft of water that had entered the school. The drainage system could not cope and all the local streams overflowed.

A man seen skulling along Clockhouse Rd in a canoe was unable to negotiate the bend because of the speed of the current, but someone put out a notice "Boats for hire". Without the Yokohama field in which to overflow, the Pool River poured over the bridge in Worsley Bridge road and covered the area in a huge sheet of water two feet deep.

In 1960, the Kent County Council formulated plans to put the rivers into culverts and so their beauty disappeared, but we still have floods, it seems at 10-year intervals. On Saturday, 1st August 1998, we were reminded that flash floods still occur when Cedars and Beckenham Roads were turned into rivers as 2 ins of rain cascaded down in 1.5 hours. Attempts are being made to return the river to its former attractive self with the landscaping at the new Sava centre on the old gasworks site at Lower Sydenham.

In 1846 when the directors of the local gas suppliers toured Sydenham and Dulwich, they decided that the area did not merit the development of gas. However in 1854, the Crystal Palace Gas Company began gas production at Bell Green, the lowest lying region in the area. This corresponded with two events that boosted sales. The first was the opening of the Crystal Palace 22.6.1854 and the second was the adoption of street lighting in 1856.

Prior to this you had to use a lantern to find your way at night and it was said that if you fell into a ditch there were so few people about you might wait days to be rescued.

The Pool River bisected the gasworks and so it disappeared into a culvert 30m below the surface to avoid interrupting the lane in the plant. When the site became the South Suburban Gas Company, nearly 2,000 people were employed there but with the advent of natural gas in the 1960's production ceased and it became redundant.

> But Oh! How changed with passing years.
> T'is now the vilest stream on earth,
> Polluted from its place of birth.
> The Kent and Surrey hills no more
> Can show their limpid rills of yore[2]
>
> Ben Jonson

Over 100 years of gas manufacture had left the soil contaminated and it was decided to move the river to a new open channel. This has been designed with a pathway linking into the Green Chain walk and to encourage nature to thrive. The gravel riverbed has already increased the shoaling of fish and the riverbanks are planted with native trees, shrubs, herbs and wetland marginals. Bridges are situated at viewpoints and imported boulders provide inviting places for children to play.

The 352 bus from Beckenham takes you to the beginning of the walk by the Sava centre but dogs are not allowed into this short stretch of the river. Access to the wild part of the river beyond the riverside walk is possible at several places, especially near the one-way system at Catford bridge where you can see the stretch once known as Gudgeon's Swim. The confluence of the rivers Pool and Ravensbourne is just above this point.

About 100 years ago, residents of the odd numbers of Catford Hill had landing stages at the bottoms of their gardens. They relaxed in their rowing boats on the river and enthusiasts held a fancy dress regatta each year. The inn on the other side of the road was called the Ravensbourne tavern until the name changes of recent times. It was first licensed in 1845.

The source of the Pool River is difficult to trace. I believe that it rises near to Norwood Junction by Werndee Rd and Shinners Close and runs along the ground water sewers below Marlow and Ravenscroft roads. In old maps it was called the Boundary stream and you can detect its course from the walls on either side of the roads along this valley. It passes along the backs of the gardens in Reddons Rd and crosses behind Cator Park School into the Park. Here it meets the flow from the Chaffinch Brook and the river Beck and from now on is called the River Pool.

THE CHAFFINCH BROOK

Climb steadily up The Glade from Long Lane or along Orchard Way from Upper Elmers End and you come to the source of the most Easterly branch of the Chaffinch, the ancient Ham Farm. Many postcards exist from the first decade of the 20th century that show the Ham Farm footpath into Elmers End.

In the 15th century, the land here was part of the Monks Orchard Estate owned by the Squery family. The estate remained intact until Baron Gwydir died in 1820. It was divided up into lots and Ham farm was sold separately from the rest to John Maberly for £10,000.

Ham is a common name for old English settlements meaning farm, homestead or estate. In earlier references, "hamm" means meadowland by a river. The ancient manor of Ham was described as an open hall, timber framed, medieval house. It was rebuilt in the 1850's and survived until 1935 although by then it had lost most of its land to the second suburb-building boom from 1925 to 1935. By 1920, Percy Harvey estate agents advertised plots for sale with no attempt at planning. Small firms and individuals with little experience built houses. There was no main sewerage or reliable water supply and such roads as existed were unpassable in winter.

By 1937, there was a population of 4,000 in some 1,500 houses but few shops and only one school and one church. This lack of foresight can be seen in the irrational road patterns of today although Orchard Avenue and The Glade are now rush hour private "motorways".

The 1850's farmhouse could still be seen to the left behind numbers 1 and 3, The Glade, before it was pulled down in 1935. When it was sold in 1921, it was listed as a property with buildings, orchard, lawn, arable land, two ponds, a stream and a spinney. The horseshoe duck pond lay to the East of the farmhouse and the East Chaffinch was the stream. It ran between Gatton Garden Mead and Park field along a line of trees to Elmers End.

With the development of the housing estate, the river was put underground partly because it was felt to be a health threat in the days before immunisation against diphtheria and antibiotics to control scarlet fever. It can be seen at the foot of the allotments near Abbots Way and Aylesford Avenue where it runs along an open culvert and meets the river that flows near Stroud Green in Shirley.

What can we see to remind us of Ham farm? Traces of the Ham Shaw (wood) remain in patches of woodland at Long Lane, the bird sanctuary, between Lorne Gardens and the Glade and along Woodland Way.

Aldersmead Avenue is the only recollection of the Alders, the old woodland by which the Chaffinch flowed. There is also a road called Ham View off Orchard Way that could be a reminder of the walk from Ham Farm. In the garden of St George's church in the Glade, there was a wooden gatepost with a plaque at its base. It was the lodge gatepost at the end of the drive to the farm (now Orchard Avenue).

Once at Elmers End few would know of the existence of the Chaffinch Brook. It runs in a culvert close to the Bollom's paint factory and passes under the Road Bridge on its way to Clockhouse.

The main source of the Chaffinch Brook is from a spring South of the Fox that feeds the lake in the Addington hills on the golf course.

Access is possible when the Scouts are using the site for their camp in the woodland called Pinewood at the top of Shirley Church Rd. There the children enjoy many activities like canoeing and crossing the water by rope bridges.

The original course of the stream can only be guessed from old maps but it surfaces briefly by Hazel Close at Shirley Oaks where all the roads in the estate have names of plants, like Primrose Lane. In the old map of Shirley, the Chaffinch appears just North of the Shirley cottage.

The stream disappears through the Ashburton playing fields where thirty years ago it could still cause serious flooding in Bywood Avenue. Now it is most of the time represented by a dry riverbed winding through the trees that once lined its banks although some water returned during the wet winter of 2000/01. This was the course of the Chaffinch with its tributary from Stroud Green.

The playing fields and adjacent areas have a long sporting history. Until well into the 19th century, a racecourse survived on the ground, which is now the site of the Woodside fire station and the Stroud Green housing estate. It is said that public horse racing at Croydon dates from the beginning of the reign of James I. Woodside station opened in 1871 bringing racing enthusiasts to the course and Queen Victoria's Jubilee on 22.6.1887 was celebrated there[3]. Eventually the mayor of Croydon closed it in 1890 because it attracted the riffraff! For 60 years the land was leased to the Beckenham Golf Club but then the Ashburton Secondary Schools were built on the site. There is still a race called the Woodside stakes at the Brighton races.

Now that the tramlink is completed, this destitute line may once more be of use but only time will show whether it can oust the car. The river shows itself again as it joins the East branch in the culvert near Elmers End.

Two more streams enter as the river travels along its culverts parallel to Clockhouse Road. When past the station at Clockhouse it runs quietly on through the allotments into Cator Park. There, our 1950's children could fish for minnows and sticklebacks at the junction with the Beck. Today, it is all tamed within concrete and no one can play.

Going back nearly 200 years, the following account was written about the Chaffinch at Elmers End.

"On our way back from Penge, W. thought this object worth sketching. He occupied himself with his pencil and I amused myself with dropping grains of dust among the fleet of tadpoles on the yellow sands. A few inches from them, in a clearer shallow, lay a shoal of sticklebacks as on the Dogger Bank. The rivulet crosses the road from a meadow where I heard it in its narrow channel. While I seated myself by the wayside among ground ivy and periwinkle, discriminating the diminutive forms of trees in the varied mosses of an old bank I recollected descriptions I had read of the gigantic vegetation on the Ohio and Mississippi. A labourer told us that this little brook is called "The Chaffinch's River and runs into the Ravensbourne".

> Ancient Charity let flow this brook
> Across the road for sheep and beggar-men
> To cool their weary feet and slake their thirst.

We must not think that all was idyllic in the past however. There were many reports in local newspapers in 1883 complaining about the footpaths. "Why has the Local Board employed roadsweepers who wore a channel sweeping the centres of paths making them all sodden and rotten? The money would be better spent gravelling the paths properly.

The path between Cedars Rd and the wheelwright's needs its hedge trimmed because the path is always wet from the incessant drip from it. The path between Yew Tree and Cedars roads is always under water because the ditch is stopped up on either side. There is also a spot in Rectory Rd just below the river that was made a soft spot extending right across the path by the drip from the trees. It was a <u>disgrace</u> to this expensive village to have paths where the unsteady villagers could have a bath at the same time as they went for a walk. All that was needed was a few loads of gravel."

The board answered some of the complaints with the usual "It would increase the rates by a third if money was spent on footpaths although the members admitted that there was no real footpath between the George and the Greyhound. They pitied the customers who fell into the holes."

On Monday, August 5th 1996, the Chaffinch at Elmers End was right back in the news.

At 3am residents were evacuated from their homes as a £1million blaze destroyed the Bollom paint factory. At least 250,000 gallons of water were used to control the blaze. A clean-up operation to remove chemicals from the rivers went on for several days. Sandbags were placed at Kingshall, Clockhouse and Moremead Roads to prevent the polluted water going further downstream. Residents at the bridge over Southend Lane had to tolerate day and night pumping of the "white" water into the sewers by the railway bridge at Bell Green. Further back the Pool River was littered with hundreds of paint cans jostling under the arches of Worsley Bridge.

This was the second disastrous fire at the works. At lunchtime on 15.9.1982, a blaze gutted the cellulose paint production building and most of the 1,000 gallons of foam used to extinguish the fire found its way into the Chaffinch.

The Chaffinch Brook and the River Beck are represented by two wavy lines on the shield of the armorial bearings of the Borough of Beckenham. Above them are chestnut trees in bloom with the white horse of Kent below.

Over the shield, the crest bears the Cator family lion and ecclesiastical symbols of Bishop Odo of Bayeux, half brother to William the Conqueror. Finally the Tudor supporters were added with the motto "not for ourselves alone" (non nobis solum). The arms were first developed in 1931.

Sadly Beckenham became a satellite of Bromley with the organisation of the London Boroughs and the arms disappeared along with the Beckenham Town Hall. (See back cover)

THE BECK

This is the best known of Beckenham's rivers because it is accessible over most of its 4 miles or so. In 1086, Bacheham, later Beckenham, was perhaps derived from the Anglo Saxon "Beohha's village". (Wallenburg's "Place names of Kent") In Wallenburg's accompanying volume "Kentish Place names" there is the following description of the river: "Beckenham is situated on a winding stream called the Beck, a tributary of the Pool River. When passing Beckenham the stream which has hitherto run North takes a North West turn. Just South of Beckenham the stream extends into a narrow lake. The modern river name is clearly a late back-formation from the name of the place"

Thus it is considered wrong that the ancient village of Beckenham gained its name from the river but vice versa. It rises in Springpark woods in West Wickham and runs down the backs of the gardens in Copse Avenue via the Alders to the White Hart. Although the White Hart pond is no longer there, the river meanders through High Broom wood, This provides a pleasant short ramble from the entrance in South Eden Park Rd.

It passes under the road to the grounds of the one-time Wellcome laboratories where the East Beck from the Park Langley golf course joins it.

Passing under South Eden Park Rd once more, the Beck flows through Harvington. This short stretch, much loved by children and dogs alike is perhaps closest to the rivers of my childhood. Onwards through Kelsey Park to lakes, bridges, waterfalls and landscaped gardens, the river provides a haven for the town's workers and a walk-through to Beckenham from Park Langley. Except for where the Beck flows between The Drive and Rectory Rd, the river has been culverted since the late 1880's. There is no public access until it joins the Chaffinch in Cator Park.

When Woodbrook School was privately owned in the 1950's, a group of girls collected a large number of horseleeches from the Beck and left them in jam jars in the laboratory. The horrified Miss Mead, headmistress, greeted me the next morning with a request to capture all the leeches that were hanging from the benches waving their tails. They had escaped during the night in search of a meal!

THE EAST BECK

The source of this little stream less than a mile long is a lake in the Park Langley golf club.

It flows concealed by trees and bushes beside the girls' school sports field to a culvert in St Dunstan's Lane. From there it enters Langley Court until recently the property of Burroughs Wellcome. Nature trails were set up through the grounds and over 3,000 hardwood trees were planted to conserve the wildlife of the region after the 1987 gale destroyed the heart of their old woodland. The East Beck supports a far wider range of plant and invertebrate life than any other river in our part of the borough since it is not polluted by land or road drainage and hardly subject to flash floods. The water is always clean and contains large shoals of minnows and sticklebacks as well as caddis flies, damsel flies and water fleas so that kingfishers regularly nest where the banks are steep.

A favourite memory is of a fox, Emma, born under the roots of a horse chestnut tree felled by the gale. She raised three litters there and lived for five years, having been a frequent visitor to the social club for titbits that she would take from the hand. Let us hope that the possible redevelopment of the area can take place without destroying its wildlife.

THE LANGLEY ESTATE

When Lord Gwydir died in 1820, his Beckenham estate, Langley, consisted of 3,202 acres.

My brother and I take to the water with the dogs

Himalayan Balsam

The map shows the eight acre field of Copers Cope farm near Lower Sydenham station. It became the Yokohama Bank Sports ground.

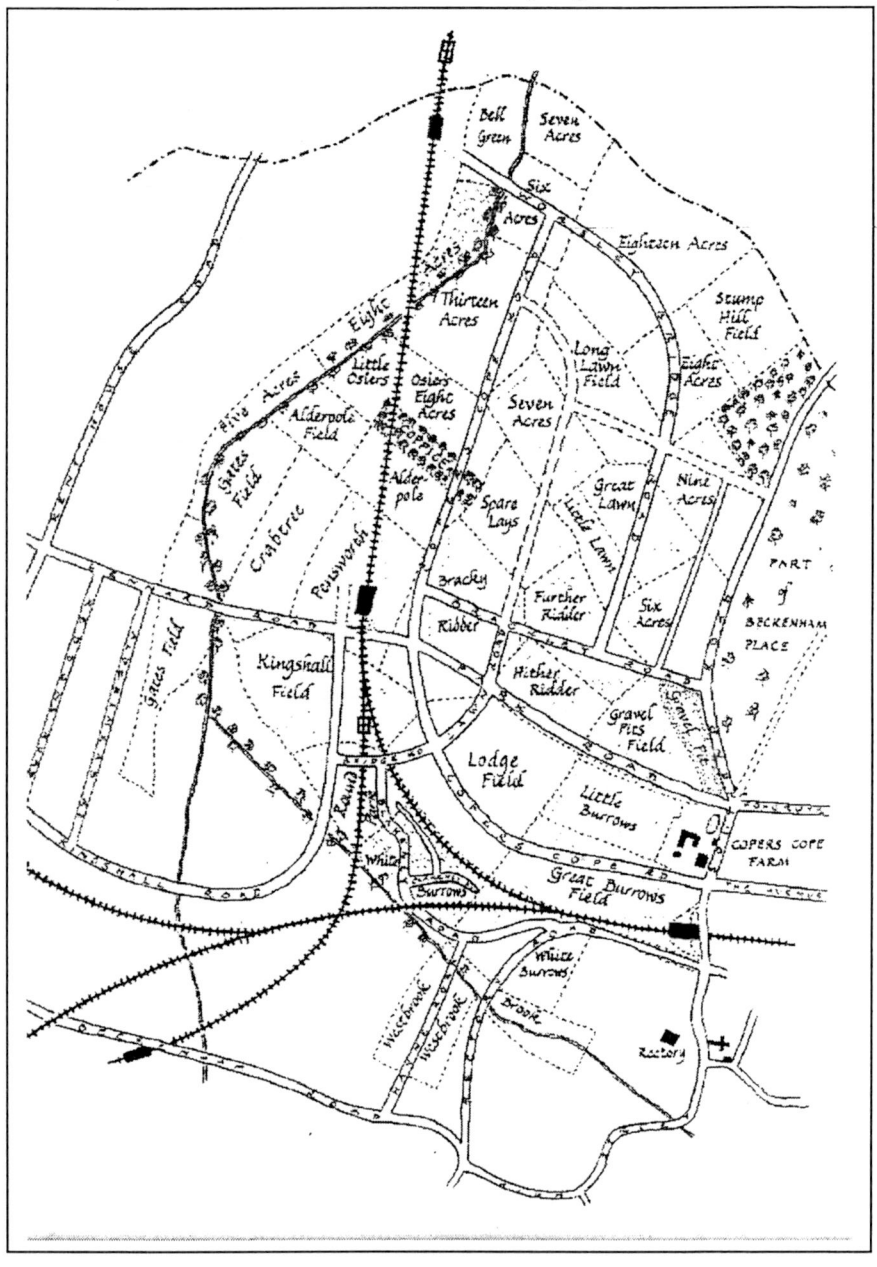

Early photographs of Lower Sydenham station

The bomb crater in the field with the author aged 12. Note the height of the poplars alongside the railway

The view in December 2000 from where the bomb fell in 1940

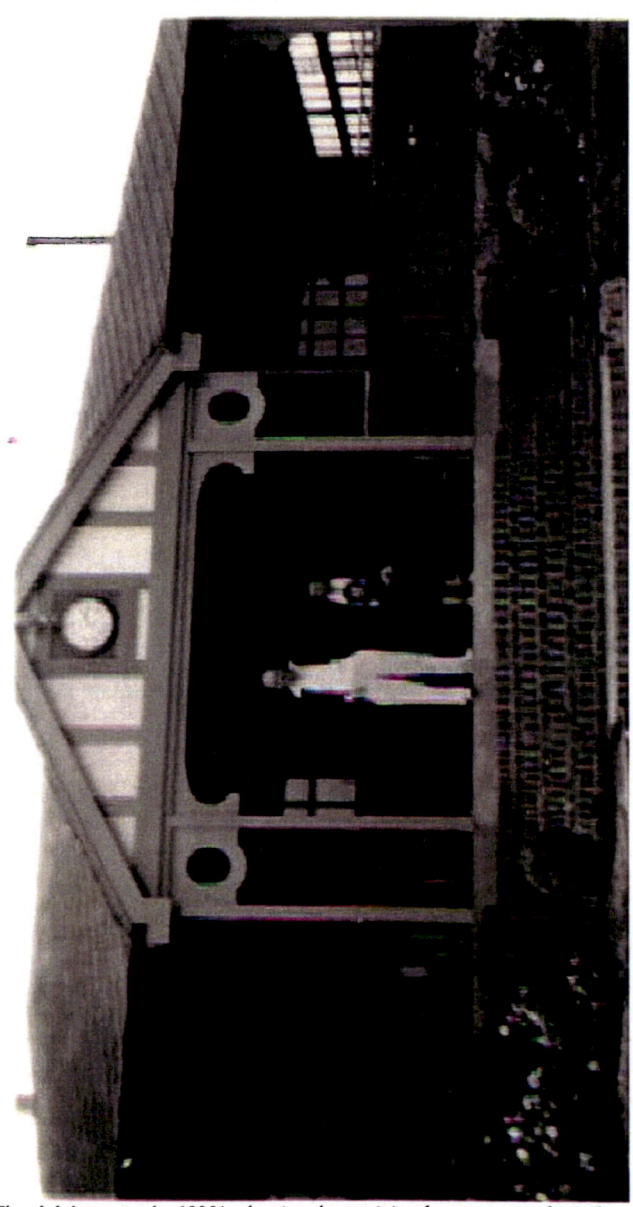

The club house in the 1930's showing the semicircular steps, now buried

The trees show the course of the river dividing our field in the foreground from the National Provincial Sports ground and Copers Cope Rd

The river Pool as it runs through the landscaped area near the Sava centre at Lower Sydenham

The pink flowered Butomus is one of the wetland marginals planted by the River Pool as it flows through landscaped area at Lower Sydenham

The junction of Worsley Bridge road with Southend Lane now--------

----------and then!

The farmhouse of Ham Farm can be seen behind the advertisement hoarding for houses and chalets

The old gatepost from Ham farm that was at the churchyard

The Chaffinch lake in the Shirley Hills

The river Chaffinch running through the housing estate at Shirley

The normally dry riverbed of the Chaffinch through the Ashburton playing fields although water was flowing in the wet winter of 2000/01

In February 2001 there was once more some water flowing in the Ashburton Chaffinch

*In this old map, the Chaffinch appears just North of the Shirley Cottage.
The lake from which it flows lies south of The Fox*

So how did the locals come to give the name Chaffinch Brook to their river? Bird catchers came to the river from London to trap the songbirds. Most of the working classes had a caged bird to brighten the weary days.

LONDON BIRDCATCHER 1827

The Chaffinch at Bollom's in Elmers End

The old bridge over the Chaffinch at Elmers End

LOT 13 An Attractive Residence
Known as
HOLLY LODGE

Situate on the South side of Upper Elmers End Rd, in the Parish of Beckenham equidistant between Elmers End and West Wickham stations with grounds of about

3a.   1r.   39p.
THE RESIDENCE

Placed on raised ground facing the road and approached by a pretty **Carriage Drive** with flower borders and lawn on eother side. It is substantially built of brick with tile roof and the **Accommodation** which is very conveniently arranged on two floors comprises

ON THE GROUND FLOOR: **Entrance Hall** with tiled fireplace and hearth; **Business Room** measuring 14ft square with fireplace and casement windows to garden; **Inner Hall** or **Lounge** measuring 20ft 6ins by 14ft with tiled fireplace and doors connecting **Drawing Room** a spacious apartment measuring 19ft by 16 ft with fireplace and casement windows to garden; and **Staircase Hall** with **Passage** connecting **Large Dining Room** measuring 26ft by 17ft with fireplace; **Separate Lavatory** with basin (hot and cold) and WC; **Kitchen** with double oven range, cylinder hot water supply, boiler and cupboards; scullery with sink (hot and cold) ;Pantry with sink and large cellarage.

ON THE UPPER FLOOR: Approached by a well-lighted staircase and corridors; Eight bed and dressing rooms, store room, bathroom with basin (hot and cold) and two WCs. Attic room above.

Company's gas and water laid on. Outside woodshed

THE GARDENS Lawns, Tennis Court, Rose bed, Large kitchen garden, Orchard, Vineries, Fowl houses

St James's stream at Upper Elmers End at the side of Eden Park school

Near the Rising Sun at Upper Elmers End where the St James's stream passed to the other side of the road

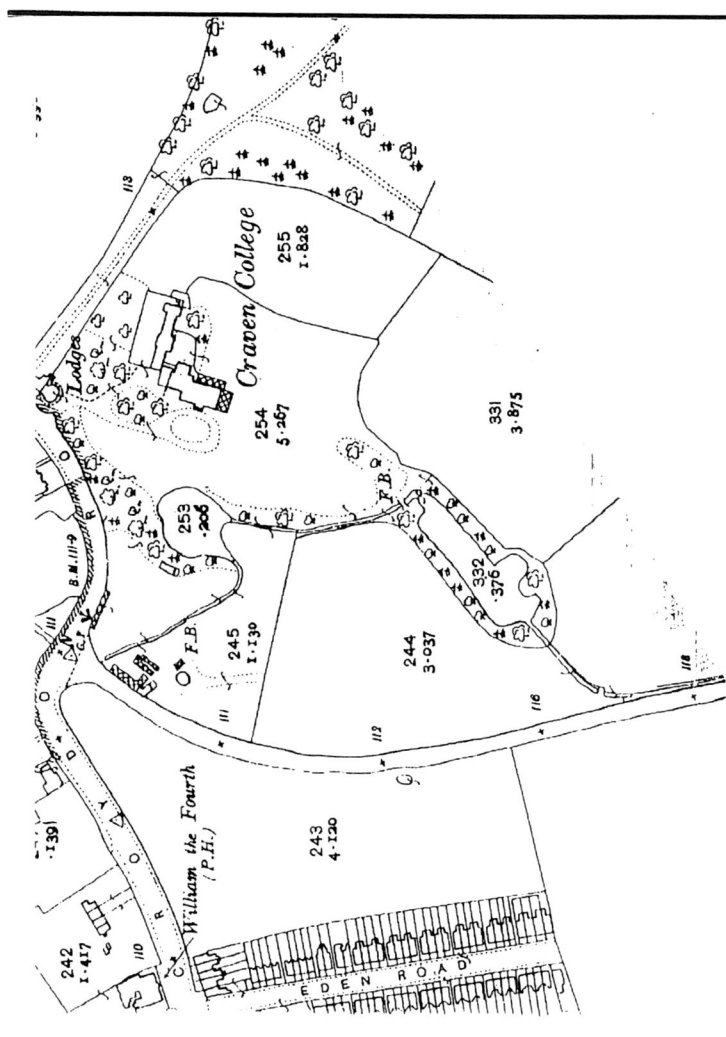

The ponds at Elmers End on the present site of Elmer Lodge and previously Craven College

EDEN PARK ESTATE
BECKENHAM

WHY LIVE IN FLATS OR HOUSES PAYING HIGH RENTS ALL YOUR LIFE?

The river Beck in Harvington

The Croydon canal at Anerley

Miller's pond

The railway line to Catford passes over the Ravenchourne

The Pool (bottom) meets the Ravensbourne (top)

Ben in the clover by the lake of the South Norwood Country Park

The lake in the SouthNorwood Country Park in January

Beckenham Church, Kent.

The wall of the Manor House gardens is on the left of the flock of sheep

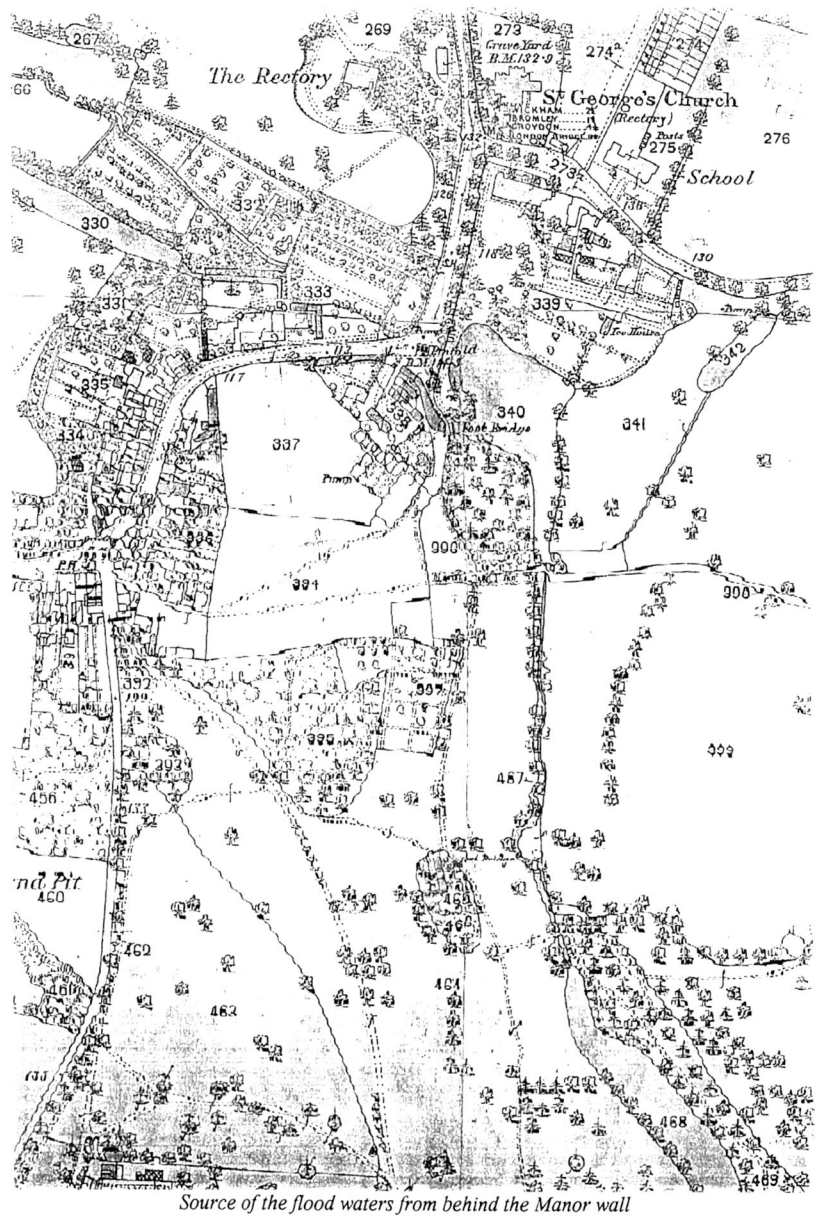

Source of the flood waters from behind the Manor wall

The flooded main street of the village of Beckenham by the Greyhound on April 10th 1878

The flooded High Street in 1968

From this Bromley & Penge advertiser picture of Jan 1949 you can see the wood behind Beckenham High St through which the Beck flows[7]

The Bourne at Corkscrew Hill has reappeared with the accumulated rainfall of 2000/2001

It stretched from Monk's Orchard and Ham farm in the South to Elmers End in the West to Beckenham High Street in the North and Bromley in the East. Thus the river system of the Pool largely rose or flowed through the estate but one needs to look hard to find where the rivers are today. The foldout sketch map (see back of book) attempts to give an idea of their present paths.

The Head girl of Langley Park School for Girls, Jane Dawson, designed a new blazer badge for the school when the former Beckenham County and subsequently the Grammar School moved from Lennard road to Park Langley. She incorporated the Arms of the Style family who bought the property in 1510 from Ralph Langley and occupied it until 1729, when Sir John Elwill, who had married Elizabeth Style, died.

There are several springs around the school and it is not uncommon to see mallard scooping their beaks through the surface water on the grass in the front after heavy rain.

THE CROYDON CANAL

The Napoleonic wars made passage by sea dangerous. Industrialists envisaged transport by canal from London to Portsmouth, which would carry building materials like stone, brick, chalk and timber in large barges.

In November 1799, Ralph Dodds produced a report that outlined a 9-mile long canal along the valley of the rivers Pool and Ravensbourne. Fortunately for "my" river valley, this scheme was set aside in favour of a more grandiose plan from John Rennie. His 10.5 miles long canal used the high ground along the ridge of Sydenham and Forest Hills and descended by long inclined planes to the Thames at Rotherhithe. So the Croydon Canal was started in 1801.

By the time it was opened, the battle of Waterloo had almost been won and there was no longer the need for internal water transport. Mill owners along the Wandle River objected to water being extracted from the Wandle's tributaries to fill the leaky canal. An engineer, Joseph Gibbs was commissioned to construct a railway.

By 1830 the canal was bought by the London and Croydon Railway company. Parts of the canal were used for the railway opened in 1839 and Penge Wharf became Penge West railway station. Anerley station was built on land given by a Scot, William Sanderson, on what had been a most attractive bend in the canal but most were pleased to see the back of a costly and often rather smelly venture. Certainly if Ralph Dodds river valley scheme had been employed, using the route taken by the Hayes to Lewisham railway today, I should not have paddled in the Pool river!

ST JAMES'S STREAM

Most of the Elmers End residents would not know of the existence of the stream that joins the Chaffinch near to St James's church because it runs in a culvert. The well-known photograph shows the Rising Sun at Upper Elmers End close to the point where the stream crossed under the road.

The 1909 OS map shows the stream flowing freely from ponds between the Elmer lodge and Upper Elmers End road. The stream rose at ponds at Spring Park farm just beyond Monk's Orchard and flowed North to Upper Elmers End by the Rising Sun. It turned West through a right angle and followed the line of Upper Elmers End road until Elmers End.

The ponds by Croydon road had previously provided a place where the floodwaters could accumulate but these became a long-lost memory as the area became built upon. The old map of Elmers End shows the boys' school, Craven College, which today is the Elmer Lodge in Dunbar Avenue. At the back there was the Eden Park Polo club where now there are the rugby and football grounds in Balmoral Avenue. The nearest to polo it reached in recent years was when the Beckenham swimming club water polo section held their annual barbecue there in July!

The stream can be seen between Eden Park School and the first of the houses where there is a gauge to keep watch on its height. It became evident after the storm in 1868 when the road in front of the Rising Sun became a lake for several days!

Elmer Lodge stands on the site of one built in 1610. It is said that the poet Shelley once lived there. In 1820, it was sold as part of the Gwydir estate and bought by Edward Richard Adams, magistrate and churchwarden at the Parish church. In 1860, the house was demolished and the present building was erected on the orders of John Goddard, insurance actuary, after whom the local road is named.

At the time the grounds still retained signs of their past beauty, with the little stream, its quaint bridges, the lake in front of the house and the medley of winding paths but the area was ripe for development. The poster shows how the local advertiser visualised this part of Beckenham as "The Garden of Eden". When I moved to Birchwood Avenue in 1971, the milkman could remember the fields before the houses were built. My house was built in 1932 at a cost of £864.[4]

MILLER'S POND

The sale of Addington Park in 1802 stated that the lessee of Spring Park Farm was responsible for maintaining the head of water and for two ponds.

Spring Park Farm was on the site used by Bethlem hospital today and the public is not permitted to see round the grounds.

Further south across the A232, there used to be more farm buildings and three ponds, the largest being Miller's pond. This was named after the last family to work the Spring Park Farm. Miller's pond may have been created and kept open to assist the farm maintain the head of water stated in the deeds and to soak cart wheels so that the rims would not dry and shrink. It was taken over by the council in 1934. The area of farm buildings became Farm Lane and one of the two smaller ponds remains at the back of Farm Drive.

Today the site is a quiet park with few amenities but plenty of water birds use the pond. Perhaps this is the true source of the St James's stream and not the springs in the Bethlem hospital parkland.

THE RAVENSBOURNE

Today the Ravensbourne rises on the 420ft contour line at Caesar's well on Keston common.

On Keston hills wells up the Ravensbourne,
A crystal rillet scarce a palm in width,
Till creeping to a bed, outspoken by art, It sheets itself across reposing there.

> Thence, crossing mead and footpath, gathering tribute,
> Wanders in Hayes and Bromley, Beckenham Vale,
> And straggling Lewisham to where Deptford Bridge
> Uprises in obeisance to its flood,
> Whence it rolls on to swell the master current
> Of the "mighty heart of England".[5]

This Ravensbourne of Ben Jonson could hardly be written of the river today although observant bus passengers may have caught a glimpse of the river during the roadworks at Bromley South in 1997.

It flows through Bromley to Glassmill Lane and along the back gardens of Ravensbourne Avenue. When it passes under Farnaby Road the level has dropped to 115ft and it was once an inviting place of fish ponds, stepping stones, rills, springs, summer houses and water walks in the gardens of Lord and Lady Farnborough.

The Bromley Record of 1.7.1878 comments that "the walk is quite impracticable to ladies because of the practice of the Bromley youths bathing in the pond at the bottom of the hill. The path is only one or two yards from the spot, which they choose and the water nowhere seems to come much above their knees.

The language that the passers-by hear is as disgusting as the other performance."

As it flows North of Beckenham Place Park the river is released from its culvert again to run at the bottom of the sports grounds on its way to Southend. Even here though the river does not flow as freely as old pictures show but has been dug down and formalised. By the time it has reached Southend and flows into Southend pond, the river has descended to the 91ft line.

Until the mid- 1920's, Southend was still a favourite spot for walking or cycling at the weekend and in the 1930's many of my generation remember the pond when it was called "Peter Pan's Pool". The island in the centre was occupied by Snow-white and the seven dwarfs, since the Plaza cinema in Catford was showing the film.

We could choose from motor boats, canoes, rowing and paddleboats. My brother and I could only go round in circles in the paddleboats because he turned his paddle so much faster than I could turn mine!

There was also a small fair which changed with the years from side shows where the aim was to catch ping-pong balls with a fishing net as they shot out of a central tube to go-carts and finally battery-driven cars. The person in charge of the playground at the back stamped the back of your hand with the date.

This mauve "badge" let us go in and out through the turnstiles.

Now Homebase, in the style of the Crystal Palace, occupies the area.

There were several mills at Southend dating from before the Norman Conquest, Batley's for grinding mustard seeds, Knapps, Livinges and Shrafholte mills and one run by John and Ephraim How, cutlers, where the Tiger's Head now stands. From this point onwards the river is canalised to go through the factory properties to Catford Bridge at the 65ft line. Immediately before receiving the Pool River, the Ravensbourne passes beneath the upper and lower level railway lines. Then from Catford Bridge it flows through Ladywell Park to its confluence with the River Quaggy at Lewisham.

THE SOUTH NORWOOD STREAM

The South Norwood Country Park lies between Elmers End station and the Beckenham cemetery. It was once much better known as the sewage farm smells from which were attributed by the Council to "the smell of fish from a passing costermonger's cart or manure in the fields".

In February 1886 the sewage farm was the subject of a public enquiry because Croydon asked for powers of compulsory purchase of land to extend it.

The evidence against the farm's extension was overwhelming. The following summary is taken from the local paper at the time.

"Residents at Woodside were concerned that their properties would be further devalued and the Beckenham Local Board agreed that the works had already retarded housing development at Elmers End. In 15 years population had only increased at Elmers End by 30 whereas elsewhere it had doubled. Over 300 complaints of the smell had been received including those from the stationmaster at Elmers End station. The manager, John Figg, said the clay soil was unsuitable because it could not absorb the sewage like the gravel at Beddington where he had worked previously. The water passed into the Chaffinch Brook and polluted the rivers as far as Lower Sydenham where the cattle drank the water and the milk pans were washed.

The Medical Officer said he regarded the farm as pernicious as it increased the outbreaks of diphtheria and scarlet fever. Contagious diseases in the area were four times greater than elsewhere. A resident of Kent House Lane said he had followed the stream from Elmers End to Clockhouse and the colour varied from light sherry to ink!" In fact still in 1960 the Pool river system downstream was nicknamed the "POO" river by the locals, myself included, but then the area's use as a sewage farm was discontinued.

Today there are two streams in the park, which converge by Elmers End Rd to form the West Chaffinch. The westerly branch is the one which on old maps bears the name South Norwood Stream but it has for a long time been confined in an ugly cracked culvert.
One benefit of the Croydon Tramlink was that part of this stream has been relocated to flow through a nature area although much of its water drains off the roads and is polluted with oil.

The easterly branch, where Jenna stands on the bridge, is more worthy of the name stream although, ironically, it has no name at all.[6] This waterway rises close to the railway line halfway between Woodside and Elmers End stations. It becomes visible at one side of the old Rylands playing field where it passes under another rustic bridge. Water is periodically allowed to flow into the water meadow on its left bank to maintain the conditions needed by the willows, tall grasses, watermint, willowherbs, thistles, fleabane, meadowsweet and many others. These in turn provide the feeding grounds for mallard, snipe, herons and others like the marsh warbler. In 1996, one pair raised two nestlings in the park.

At a recent count, nearly 300 herbaceous plants, more than 154 bird species, 28 butterflies and well over 100 moths had been identified in the area including the striking elephant hawk moth.

This stream also maintains the water level of the lake with a supply under another rustic bridge nearer to Elmers End Rd. The river provides local children with the opportunity to fish for sticklebacks and tadpoles and a pier was constructed to assist their investigations. In both 1998 and 2000, a pair of kingfishers nested in the bank of the lake and raised 2 youngsters. The adults could be seen plunge diving to catch fish for their brood. They would perch on a bare branch and turn so that the sun alternately reflected off their chestnut breasts and turquoise wings.
When the Tramlink was being constructed through the Park, I was concerned for its effect on the land. Once more the earth was suffering to accommodate our needs and desires.

ON THE HILL

As I stood on the top of the hill
And looked down over the gentle green slopes below,
I imagined that I could see still
The cottage gardens of the workers long ago.

For a century or more they say
A sewage farm was based on the land before me,
With spreading fields of rye and hay
A stable, barn, ploughs and other machinery.

By the 1960's the area grew wild,
Still with rhubarb and gooseberries in old gardens over-grown.
The land was no longer defiled
But left to butterflies, plants and animals on their own.

And for close on a decade
We have enjoyed the scents, bright colours and sound,
For a lull in the parade
Of insults man confers on the fields around.

Behind me now as I stand on the hill,
The ground has been savaged and sliced by the rails.
And by next year we know that there will
Be trams taking people to Croydon's ceaseless sales.

PICTURES OF COUNTRY LIFE (1847)
In an old book by Thomas Miller, a summer ramble is described from Anerley to Beckenham. "Having had a glass of ale and a crust of bread and cheese at the Woodman, we will strike down the hill and peep at Annerley(sic) station. We shall have woods on either side. There runs a rabbit! That was a pheasant, which sprung up before us! There's woodbine for you, you might gather an armful.

What a variety of beautiful flowers are spread at our feet! This is a place where the inhabitants of London come in hundreds on a Sunday to breathe the fresh air, for once in Croydon railway carriages they are wafted here in a few minutes. Ten years ago it was wild woodland. Where the Croydon and Brighton railway now bears its brown iron track there stood an old canal. You could walk along the banks between the woods and see the shadows of the trees and the deep blue sky reflected in the water and behold hundreds of beautiful flowers bending over and looking into the bright mirror which threw back their images.

Straight down the road will lead us to Beckenham. The scenery on either hand is very beautiful. At the end of those cottages is a real green lane. It leads nowhere except into the fields. Excepting in the hay season or to carry in a load of manure, not a wheel crushes down the grass with which the almost untrodden lane is carpeted. It is full of elbows and goes winding here and there like the brook whose course it follows.This is the entrance to Beckenham.

What a splendid old manor house with its high crumbling wall, overgrown with moss and lichen! Read the monuments in Beckenham church and you will find that many a distinguished man ended his days in this ancient village."

Before 1878, after the R.Beck left Kelsey Park it flowed into a large lake in the grounds of the Manor House described by Thomas Miller. In the sketch, the wall is on the left of the flock of sheep coming down Church Hill.

THE END OF THE WALL

The Manor house was opposite St George's church and the river Beck fed the lake at the bottom of the hill. "After a deluge for two days and nights in April 1878, debris blocked the outlet grid in the manor grounds and stopped the outflow of water until the level reached the top of the boundary wall. On the morning of the 11th, 120ft of the 15ft high wall collapsed with the sound of thunder and a sort of tidal wave rolled down the street and filled the lower part of the village with water. Everything was washed out of the shops and old ladies had to be rescued from upper windows. The rustic bridge came floating down the stream and the writer of this account (written in the local paper in 1928) had some fun retrieving it and navigating the High St to Christchurch." The map shows Beckenham with its lake contained behind the wall, but it is not surprising that the High St is so susceptible to flooding. A few days before the picture of the Greyhound was published in the local paper the same part of the High St was flooded after a 1.5hr downpour in the morning.

This was just as commuters were trying to get to work. As usual, Cedars Rd opposite the hospital was turned into a lake and Elmers End station could not be reached because of the flooding of the Chaffinch. Cyclists had water up to their wheel hubs and New Beckenham station was unapproachable as the Pool flooded.

Of course our settlements naturally grew round the rivers and their easily fordable crossing points where transport routes converged, such as at Catford. However the floodplains of the rivers and the channels through which they flow have become very restricted by the growth of our towns. It has been apparent at least since 1878 that Beckenham has problems with flash floods. Filling in the lakes at Elmers End and Beckenham and the building up of the ground at Lower Sydenham increased the trouble. Today the Environment Agency whose aim has been to restore the flood plains where possible manages the rivers. The most important of their schemes has been the improved flow through the Riverside Walk at Lower Sydenham. By purchasing the land, J. Sainsbury enabled the BG Property Division to use the services of the landscape artist, Marie Burns, to construct a new path for the river Pool that can accommodate at least some of the river's floodwaters.

ACKNOWLEDGEMENTS

1. Maps of the Cator Estate and others from Bromley, Beckenham and Shirley libraries, together with willing help from the local library staff.
2. Archives from Lewisham library.
3. Hone's Table Book vol I.
4. Local newspapers on microfilm at Bromley library.
5. Archives at Shirley library and Anerley library.
6. Jim Linwood of the Environmental Agency.
7. Many friends who passed on their reminiscences, especially Margaret Watson and Pat Hollanby.
8. Gill Dunn who traced the river Pool with me at Catford Bridge.
9. My brother, who had a better memory than mine.
10. The Scouts and their leaders at Pinewood Lake.
11. Olive, who told me of the spring that had to be put into an underground culvert when Shinners Close was built by Norwood Junction, thus suggesting a source for the Pool river.
12. Special thanks to John Bedford for proof reading the copy.

REFERENCES

[1] For those interested in the Roman road, see Ivan Margany's "Roman Ways in the Weald" pp 126-130, which uses the reports of excavations by B.F.Davis "From West Wickham to London" from the Surrey Archaeological Collections, 43,61. Also "Roman roads in NW Kent" by Paul Walters, reviewed in Bromleage Dec 2000

[2] Nathaniel Daw's "History of Deptford"

[3] See "The Croydon Races" in Norwood Rev. Nov 00.

[4] See the copy of the advertisement for the sale of the residence Holly Lodge in centre pages. It became the school now known as Eden Park School although pupils in 1943 remember it as the Holly Lodge School under Mrs Mallick.

[5] Hone's Table Book 1837

[6] It was a casual query "Do you know the name of this river" that started my thoughts on Beckenham's rivers. I am still unable to answer the question!

[7] Bromley & Penge Advertiser Jan 1949. Owners of houses in The Drive and Church Ave banded together to buy the land in the heart of the town. The church at the left of the picture is Christchurch.

Anerley & Croydon canal	20,21,32
Beckenham Arms	16,17
Bell Green	3,7,16
Bollom's paint factory	12,16
Boundary stream	9
Cedars Rd	7,15,34
Clockhouse station	5,6
Copers Cope Farm	2,4
Copers Cope Rd	3,5
Craven college, Elmer Lodge	22,23
Eden park school	23
Garden of Eden	23
Gas Companies	8
Greyhound	15,23
Gudgeon's swim	9
Ham Farm	12-Oct
Jane Dawson	20
John Cator, timber merchant	3
London to Lewes Roman Rd	4
Lower Sydenham station	3
Manor House	32,33
Miller's pond	23,24
National Provincial sports ground	3,5
Pavilion now Footzie club	5
Racecourse	13
Rising Sun	22
River Beck	6,9,16-19,33
River Chaffinch	5,6,9-16, 18,28,34
River Pool	1,2,4,5,8,9,16,17,20,21,28,34
River Ravensbourne	14,21,24-27
River Wandle	21
Rover	1,5
S Norwood Country Park	27-31
Sava centre	9
South Norwood stream	27,29
Southend pond (Peter Pan's Pool)	26,27
St James's stream	22,23
Stroud Green	11,13
Thomas Meadows	5
Unnamed stream	29,30
Woodbrook school	18
Worsley Bridge Rd	3,6,16
Yokohama Sports ground	2,3,5